Andréa Meiado Chiarioni
Marcel R. da Silva

Using Solid Waste in Environmental Education

Andréa Meiado Chiarioni
Marcel R. da Silva

Using Solid Waste in Environmental Education

ScienciaScripts

Imprint

Cover image: www.ingimage.com

This book is a translation from the original published under ISBN 978-620-2-04237-6.

Publisher:
Sciencia Scripts
is a trademark of
Dodo Books Indian Ocean Ltd. and OmniScriptum S.R.L publishing group

120 High Road, East Finchley, London, N2 9ED, United Kingdom
Str. Armeneasca 28/1, office 1, Chisinau MD-2012, Republic of Moldova, Europe
Managing Directors: Ieva Konstantinova, Victoria Ursu
info@omniscriptum.com

Printed at: see last page
ISBN: 978-620-8-52758-7

Index

USE OF SOLID WASTE IN ENVIRONMENTAL EDUCATION ENVIRONMENTAL EDUCATION: USE OF SCHOOL CHALK WASTE AS A IN ORNAMENTAL PLANTS.

ANDREA MEIADO CHIARIONI[1]

MARCEL RICARDO DA SILVA

ABSTRACT: The aim of this work is to demonstrate, through a practical lesson, the importance of reusing solid waste, in this case school chalk waste, for students in the 1st year of secondary school at Fundapao Casa (CASA Arapa) in the city of Arapatuba/SP. Environmental education is the main instrument for transformation and is fundamental for developing a critical awareness of the environment, generating commitment and responsibility among the population and establishing a commitment to future generations. Leftover chalk, a material rich in calcium carbonate and calcium sulphate, is discarded at the end of school lessons. The purpose of the practical lesson was to use the waste from school chalk to correct soil acidity. Soil was collected from the premises of Fundapao Casa (CASA Arapa) and treated with ground school chalk waste, compared to untreated soil, in order to analyze the level of acidity in the proposed situations. Improvements in soil acidity were observed after treatment with school chalk waste for planting ornamental plants. Demonstrating to the students that environmental education brings new perspectives to solve the problems associated with solid waste, from generation, collection, transportation to reuse in new processes.

KEYWORDS: Solid waste, environmental education, school chalk.

[1] Specialization in Environmental Education from the Sao Luis School of Education. Degree in Chemistry. Technology in Biofuels from Fatec de Arapatuba. Author's e-mail: .andreameiado@professor.educacao.sp.gov.br

1. INTRODUCTION

Environmental issues are increasingly being discussed in various areas of knowledge. The training of professionals and the university community in an interdisciplinary perspective to make the population aware of the need to strengthen the interrelationships between the natural and social environments (JACOBI, 2003).

The educational area includes the prerogatives of environmental education, through which the environmentally educated community becomes aware of its global reality, the type of relationships that people establish between themselves and with nature, the problems arising from these relationships and their root causes. It develops, through a practice that links the student with the community, values and attitudes that promote behavior aimed at overcoming this reality, both in its natural and social aspects, developing in the student the skills and attitudes necessary for transformation (AMBIENTE BRASIL, 2008).

According to Conama Resolution No. 5 in its first article, the waste that constitutes solid waste is that generated in ports, airports, railroad and road terminals and establishments providing health services. This resolution serves as a parameter by defining solid waste as: "Waste in a solid or semi-solid state, which results from the activities of the community of origin: industrial, domestic, hospital, commercial, agricultural, service and sweeping" (MAZZER, CAVALCANTI, 2004).

Chalk is discarded every day at the end of lessons in schools. One property derived from school chalk is gypsum. According to (Oliveira *et al.,* 2005.) gypsum has been used as a soil acidity corrector on the European continent since the early 18th century. It is an abundant mineral in nature, composed mainly of calcium sulphate and calcium carbonate.

1.1. General Objective

- Through the learning situation of the use of lime in agriculture, demonstrate to the high school students (1st Year) of Fundapao Casa (CASA Arapa) the use of solid school chalk residue in the correction of soil for planting ornamental plants.

1.2. Specific Objective

- To make students aware of the importance of environmental education for the reuse of certain solid waste as a strategy to benefit the environment, promoting ecological sustainability.

1.3. Practice Description

The educational practice was carried out at the Fundapao Casa (CASA Arapa) located in the city of Arapatuba in the state of Sao Paulo with the support of the State School Jose Augusto Lopes Borges. The Arapatuba Directorate of Education, through its team of Pedagogical Nucleus Coordinating Teachers (PCNP), provided all the encouragement needed to carry out the practical classes, especially the technical guidance (O.T) with the mediation of the PCNP's Dorislei Aparecida Teixeira de Carvalho (Science) and Angela Maria de Oliveira Barbosa Nunes (Chemistry).

The work teams involved were Professor Andrea Meiado Chiarioni (Chemistry/Sciences) with the help of Professor Marcel Ricardo da Silva (Chemistry/Sciences/Biology) to accompany and mediate the practical class.

The teaching team at Fundapao Casa (CASA Arapa) provided

the support and structure to carry out the practical procedure at the Teaching Unit.

The student J. F. S. S. was assigned to collect the soil and prepare the school chalk waste in order to start the practical and demonstrate the procedures and results of the practical lesson to the other students.

2. THEORETICAL BASIS

2.1 History of environmental education in the world

Environmental education emerged in the second half of the 19th century. In 1864, the book "Man and Nature", or "Physical Geography Modified by the Action of Man", was released by the American Georges Perkins Marsh. Five years later, the term "ecology" was proposed by Ernst Haeckel to define the studies to be carried out on the relationships between species and their environment (ARAUJO, 2007).

At international level, the publication of the "Study on the Protection of Nature in the World" in 1951, organized by the International Union for the Conservation of Nature - IUCN, which had been created as a result of the Fontainbleau International Conference in France in 1948 (ARAUJO, 2007).

The introduction of environmental issues was marked by social struggles that began between the 1970s and 1980s. At the height of the social struggles organized worldwide, movements to defend ecology and the environment began, the landmark of which was the publication of the book "Silent Spring" (1962) by the American Raquel Carson.

Following this publication, which had repercussions throughout the world, activists from environmental movements and the United Nations Organization (UNO) held several international events that addressed the issue of environmental preservation and education (GALLI, 2007).

The first event was the United Nations Conference on the Environment (1972), known as the Stockholm Conference. With the participation of 113 countries, this event, which denounced the devastation of nature that was taking place at the time, decided that human growth needed to be rethought immediately (QUINTINO, 2013).

The main recommendation of this conference was that emphasis should be placed on environmental education as a way of criticizing and combating the environmental problems that existed at the time. Two documents were drawn up at this meeting: the "Declaration on the Human Environment" and the "World Action Plan" (ARAUJO, 2007).

The first Belgrade Conference was held in 1975 with the participation of researchers and scientists from 65 countries. This meeting resulted in a document called the "Belgrade Charter", which advocated a new ethic to promote the eradication of poverty, illiteracy, hunger, pollution, exploitation and all forms of human domination. Another important deliberation of this conference was the drafting of the principles and guidelines for the international environmental education program, which was continuous and multidisciplinary, taking into account regional differences and national interests (QUINTINO, 2013).

2.2. History of environmental education in Brazil

The Stockholm Conference influenced the creation of the Special Secretariat for the Environment. The world scenario in relation to socio-environmental issues was undergoing major debates with a view to changing panoramas, and with this in mind, the government established the objectives of implementing integrated management of environmental issues with the Special Secretariat (JANKE, 2012).

According to Quintino (2013 *apud* PERINI, 1998), the action plan for this conference suggested teacher training, as well as an action methodology for environmental education worldwide. With this policy in mind, three more international conferences on environmental education were held between the 1970s and 1980s.

For many experts, 1997 was the Year of Environmental Education in Brazil. Two facts can help to understand the rise of the theme, which made the year a milestone in the history of Brazilian Environmental Education. On the one hand, there was the commemoration of the two decades since the Tbilisi Conference. Promoted by UNESCO in 1977, in the capital of Georgia, former Soviet Union, it resulted in a final document that is the basis for the modern vision of environmental education (MMA, 2009).

The process of institutionalizing Environmental Education in the Brazilian federal government began in 1973 with the creation of the Special Secretariat for the Environment (Sema), linked to the Presidency of the Republic. Another step in the institutionalization of Environmental Education was taken in 1981, with the National Environmental Policy (PNMA), which established, in the legislative sphere, the need to include Environmental Education in all levels of education, including community education, with the aim of enabling the community to actively participate in defending the environment (PRS, 2013).

The Brazilian Charter for Environmental Education was

established during the Conference known as Rio 92, with the participation of the Ministry of Education and Culture (MEC). It recognized that Environmental Education is one of the most important instruments for making sustainability viable as a strategy for the survival of the planet and, consequently, for improving the quality of human life (BRASIL, 2007).

The National Curriculum Parameters (PCN) were approved by the National Education Council. The PCN are a subsidy to support the school in drawing up its educational project, inserting procedures, attitudes and values into school life, as well as the need to deal with some urgent social themes, of national scope, called transversal themes: the environment, ethics, cultural plurality, sexual orientation, work and consumption, with the possibility of schools and/or communities choosing others of relevant importance to their reality (BRASIL, 1997).

The integration of multidisciplinary support took place with the creation of the Secretariat for Continuing Education, Literacy and Diversity (SECAD) in 2004. Integrating socio-environmental support into the state and municipal education networks, SECAD has integrated actions in the areas of Diversity, Indigenous School Education and Education in the Countryside, thus giving Environmental Education greater visibility and highlighting its cross-cutting role (PRS, 2013).

Currently, Brazil, together with the other countries of Latin America and the Caribbean, has made international commitments with the implementation of the Latin American and Caribbean Program for Environmental Education (Placea10) and the Andean-Amazonian Plan for Environmental Communication and Education (Panacea), which include the Ministries of the Environment and Education of the countries.

Environmental Education in Brazil was consolidated through the National Curriculum Parameters (PCNs), where it is presented as one of the cross-cutting themes, as can be seen in the publication Inclusion of Science in Saeb: basic document (INEP, 2013), and should be worked on in such a way as to highlight social, political, economic and ecological aspects. Thus, as a cross-cutting theme, Environmental Education should be present in all subjects, permeating their content in order to build a more integrated vision and promote understanding of socio-environmental issues as a whole.

2.3. Solid waste

By definition, solid waste is discarded material, substances, objects or goods resulting from human activities in society, the final destination of which is solid or semi-solid, as well as gases contained in containers and liquids whose particularities make their discharge into the public sewage system or bodies of water unfeasible, or require technically or economically unfeasible solutions in view of the best available technology (MMA, 2012).

Due to the demographic growth of the world's population and the concentration in urban areas, this has resulted in an increase in the use of natural resources, both through use for production and consumption, and through the damage caused by the return of waste to nature after its use by man (GODECKE *et al.,* 2012).

To this end, in 2011 the process of drawing up the National Solid

Waste Plan began. This is one of the most important instruments of the National Policy, as it identifies the problems of the various types of waste generated, the management alternatives that can be implemented, indicating target plans, programs and support for positive changes in the current situation (MMA, 2012).

The total generation of Urban Solid Waste in Brazil in 2014 was approximately 78.6 million tons, which represents an increase of 2.9% from one year to the next, higher than the population growth rate in the country during the period, which was 0.9% (ABRELPE, 2014).

This waste, when disposed of incorrectly, accumulates and causes contamination of the water table, attracting insects and disease vectors and compromising human health (EL-DEIR, 2014).

Environmental education plays a fundamental role in showing students the problem of solid waste. The search for information on how to dispose of it should be encouraged through reading texts, articles, newspapers and support proposed in the learning situation, in order to find sustainable alternatives to minimize environmental impacts, promoting the student's role in environmental issues (ABRANTES *et al.,* 2014).

According to Schalch *et al,* (2002), solid waste is classified according to its origin, as follows: Urban: includes household waste generated in homes, commercial waste produced in offices, stores, hotels, supermarkets, restaurants and other similar establishments, service waste from urban public cleaning, as well as waste from sweeping public roads, cleaning galleries, land, streams, beaches, fairs, pruning, weeding.

Industrial waste: corresponds to waste generated in various

types of processing industries. The following grouping is proposed by ABNT-NBR 10.004 (1987), based on the hazardous nature of some of these wastes.

Class I (hazardous) waste: due to its flammability, corrosiveness, reactivity, toxicity and pathogenicity, it may present a risk to public health, causing or contributing to increased mortality or having adverse effects on the environment, when handled or disposed of inappropriately.

Class II waste (non-inert): this class includes potentially biodegradable or combustible waste.

Class III waste (inert): this class includes waste considered inert and non-combustible.

Health service waste: waste produced in hospitals, medical and veterinary clinics, clinical analysis laboratories, pharmacies, health centers, dental consultancies and other similar establishments. This waste can be grouped into two distinct levels.

Common waste includes the remains of food, paper and wrappers.

Septic waste: made up of leftovers from operating rooms, isolation areas and hemodialysis centers. Their handling (packaging, collection, transportation, treatment and final disposal) requires special attention, due to the potential public health risk they can pose.

Waste from ports, airports, road and rail terminals: this constitutes septic waste, which can contain pathogenic organisms, such as: personal hygiene and grooming materials, food waste, and vehicles from other cities, states and countries.

Agricultural waste: this is waste from agricultural and livestock activities, such as fertilizer packaging, pesticides, feed, crop residues and animal manure. The greatest concern at the moment is agrochemical packaging, due to its high degree of toxicity, which is the subject of specific legislation.

Rubble: this consists of construction waste: demolition, building site waste, excavation soil.

Radioactive waste (atomic waste): waste from nuclear fuels. Its management is the exclusive responsibility of CNEN - the National Nuclear Energy Commission.

2.3.1 Solid waste of organic origin

Organic waste is made up of animal or plant remains discarded from human activities. They are materials that, in balanced natural environments, degrade spontaneously and recycle nutrients in natural processes. But when derived from human activities, especially in urban environments, they can become a serious environmental problem, due to the large volume generated and the inappropriate places where they are stored or disposed of (MMA, 2016).

Improper disposal of organic waste generates slurry, methane emissions into the atmosphere and encourages the proliferation of disease vectors. It is therefore necessary to adopt appropriate methods for managing and treating these large volumes of waste, so that the organic material present is stabilized and can fulfil its natural role of fertilizing the soil.

Article 36, point V of the National Solid Waste Policy (Law 12.305/2010) stipulates that service providers must implement "composting systems for organic solid waste and coordinate with economic and social agents ways of using the compost produced". In this way, it is understood that promoting composting of the organic fraction of waste, as well as implementing selective collection and environmentally appropriate final disposal of waste, is part of the list of municipal obligations established by Law 12.305/2010.

According to the definitions of recycling and waste in the PNRS (Art. 3, items XIV and XV), it can also be concluded that processes that promote the transformation of organic waste into compost and fertilizers (such as composting) can also be understood as recycling processes. In this way, organic waste should not be considered indiscriminately as waste, and efforts to promote its recycling should be part of waste management strategies on any scale (household, community, institutional, industrial, municipal).

The main national legal references currently in force applying to the recycling of organic waste are listed below:

Law No. 6894, of December 16, 1980. Provides for the inspection and supervision of the production and trade of fertilizers, correctives, inoculants, stimulants or biofertilizers, remineralizers and substrates for plants, intended for agriculture, and makes other provisions. (Edited by Law No. 12.890 of 2013).

Decree No. 4.954, of January 14, 2004. Amends the Annex to Decree no. 4.954, of January 14, 2004, which approves the Regulation of Law no. 6.894, of December 16, 1980, which provides for the inspection and supervision of the production and trade of fertilizers, correctives, inoculants, or biofertilizers, remineralizers and substrates for plants intended for agriculture. (Edited by Decree no. 8.384, of 2014)

CONAMA Resolution n. 375, of August 29, 2006. Defines criteria and procedures for the agricultural use of sewage sludge generated in sanitary sewage treatment plants and its derived products, and makes other provisions.

SDA Normative Instruction No. 25, of July 23, 2009, of the Ministry of Agriculture, Livestock and Supply. Approves the rules on the specifications and guarantees, tolerances, registration, packaging and labeling of simple, mixed, compound, organomineral and biofertilizer organic fertilizers intended for agriculture.

SDA Normative Instruction No. 27, of June 5, 2006, of the Ministry of Agriculture, Livestock and Supply. Provides for the import or sale, for production, of fertilizers, correctives, inoculants and biofertilizers.

GM Normative Instruction No. 46, of October 6, 2011, of the Ministry of Agriculture, Livestock and Supply. Establishes the Technical Regulations for Organic Animal and Plant Production Systems.

GM Normative Instruction No. 53, of October 23, 2013, of the Ministry of Agriculture, Livestock and Supply. Establishes provisions and criteria for the inspection and supervision of fertilizers, correctives, inoculants, biofertilizers and secondary materials; the accreditation of private research institutions; and minimum requirements for the assessment of feasibility and agronomic efficiency and the preparation of the technical-scientific report for the purpose of registering fertilizers, correctives and biofertilizers as new products.

Studies carried out by Gomes (2004), using waste from machining processes, contribute to the reduction of storage areas, lower handling costs and the reduction of environmental pollution; and for the factory, a reduction in production costs and greater efficiency in the use of raw materials. For use as energy inputs in processes, the waste saves on the purchase of energy used by the industry in production. These alternatives to waste add value to the leftover wood and can increase the company's income.

The use of organic household waste through composting, the authors Wangen and Freitas (2010) show in their studies that household composting proved to be viable for the cycling of solid organic household waste, and in a period of 120 days gave rise to a compost with good physical and chemical characteristics, with potential for agricultural use, as a soil conditioner and/or as a substrate for plants.

2.3.2 Solid waste of mineral origin

In mining activities, large volumes of material are extracted and moved. The amount of waste generated by the activity depends on the process used to extract the ore, the concentration of the mineral substance stored in the matrix rock and the location of the deposit in relation to the surface (SILVA *et al.,* 2011).

The main sources of degradation in mining activities are the improper disposal of tailings from the beneficiation process and the disposal of materials.

Among the minerals exploited is graphite, a lubricating material characterized as a natural or synthetic material. Natural graphite is derived from mining. Refined graphite contains carbon, sulphur, SiO_2 and volcanic dust. The quality of graphite crystallization is determined by the higher carbon content, improving lubricity and resistance to oxidation. Synthetic graphite is synthesized at a high temperature, maintaining a very high carbon content of up to 99.5 - 99.9%, which is why it works very well as a lubricant (IMPLASTEC, 2011).

Graphite is an element made up essentially of carbon, which is why its color is predominantly black or very dark grey. It is widely used in Brazilian schools in pencils, pencil cases and compasses (IMPLASTEC, 2011).

Like all school materials, after use, the act of pointing causes the size of the pencil to shrink to the point where it becomes undesirable to use and dispose of. In many industries, graphite is used as a type of lubricant, as its lamellar structure in the form of dry powder makes it ideal for reducing friction between mechanical parts (MENTE CRIATIVA, 2010).

Work carried out by (Rosa *et al.,* 2017), collected all school pencils that were destined for disposal at Fundapao CASA Arapatuba during twenty school days, stored them and weighed them. The result was 500g (grams) of raw waste. Over the course of 200 school days, this amounted to 5 kg (kilograms) of waste. After separating the graphite from the wood, 87g (grams) of material was obtained, which at the end of the year amounts to 870g (grams). The study showed that it is possible to reuse the ground graphite waste from school pencil rejects as a lubricant for hinges, giving a sustainable destination to a material that would otherwise go to waste.

Limestone is used as a raw material in construction, in the manufacture of lime and cement, and is of great importance in agriculture as a corrective for acidic soils. It is also used in the paper, plastic, chemical, steel, glass and refractory industries (SILVA, 2009).

The basic composition of school chalk is a combination of limestone and gypsum. Basically, chalk is a base (an alkaline compound that neutralizes acids) made of calcium and oxygen combined with carbon ($CaCO_3$), while gypsum is a salt (a product of the neutralization reaction between a base and an acid), composed of calcium and oxygen combined with sulphur ($CaSO_4$) (GIZMODO BRASIL, 2015).

School chalk waste is generated in the study environment when the material is used on the blackboard and also when the chalk reaches a minimum size, making it unsuitable for use.

2.4 Liming

Most Brazilian soils are considered to have a low nutritional level and high acidity, which hampers the development of Brazil's main crops (OLIVEIRA *et al.,* 2005).

The soil is corrected by means of chemical reactions of calcium and magnesium carbonates, when solubilized in water, neutralizing the active acidity of the soil expressed by its pH value, this technique is called liming (PEREZ, 2007).

The active acidity of a soil is due to the presence of free H ions$^+$ generated by acidic substances contained in the soil (nitrogen fertilizers, organic acids, etc.). Its neutralization occurs when the OH^- anion neutralizes the protons, i.e. the soil improvers must contain basic components to generate OH^- and promote neutralization (ALCARDE, 2003). There are various materials with neutralizing properties, such as:

a) Limestone: originates from sedimentary rocks containing minerals with more than 30% calcium carbonate (arogonite and calcite). When the predominant mineral is dolomite.

The color of limestone varies from white to black, light and dark gray. Many limestones have reddish, yellow, blue or green tones, depending on the type of impurities they have. They are mainly found in the marine environment.

b) agricultural quicklime: a product obtained industrially by calcining or completely burning limestone. Its constituents are calcium oxide (CaO) and magnesium oxide (MgO), and it is presented as a fine powder.

c) agricultural hydrated lime: a product obtained industrially by hydrating virgin lime. Its constituents are calcium hydroxide Ca(OH)2 and magnesium hydroxide $Mg(OH_{)(2)}$ and it also comes in the form of fine powder.

d) calcined limestone: product obtained industrially by partially calcining limestone. Its constituents are undecomposed CaCO3 and MgCO3.

limestone, CaO and MgO and also $Ca(OH_{)(2)}$ and $Mg(OH)(_{2)}$ resulting from

hydration of oxides by moisture in the air. It comes in the form of a fine powder. Its neutralizing action is due to the strong base OH^- and the weak base CO_3^{-2}.

e) basic steel slag: by-product of the iron and steel industry. Its constituents are calcium silicate (CaSiO3) and magnesium silicate $(MgSiO^{)(-3})$

f) calcium carbonate (CaCO3): a product obtained by grinding marl (terrestrial deposits of calcium carbonate), coral and sambaquis (marine deposits of calcium carbonate, also known as marine limestone). Its neutralizing action is similar to that of calcium carbonate in limestone.

Liming is an agricultural practice whose benefits include correcting soil acidity, neutralizing toxic aluminium and manganese, increasing the availability of calcium, phosphorus and magnesium, improving soil structure and m*icrobial act*ivity (ROSSETO et al., 2004), as well as increasing *the ad*sorption of potassium ions (K *),* an important element for sugarcane crops, 2004), as well as increasing the adsorption of potassium ions (K^+), an important element for sugarcane crops, as it prevents its loss through leaching processes (ROSSETO *et al.,* 2008; BITTENCOURT; SAKAI, 1975).

Materials such as quicklime, slaked lime, calcined limestone, ground seashells, ash and limestone have the potential to be used as a soil acidity corrector, provided that the material is of neutralizing quality and, above all, the farmer has an economic advantage (OLIVEIRA *et al*., 2005).

(2012) used marble sawdust residues as a soil acidity corrector. The marble waste sawn with a diamond blade (LD) and with an *auger* blade (LA) can be classified as dolomitic and has the characteristics of *filler* limestone, since it has a grain size finer than 0.30 mm and reactivity equal to or greater than 100%. Both the application of the residues and the commercial limestone increased the Ca and Mg concentrations and the pH of the soil, eliminating Al toxicity. Therefore, all the materials tested have adequate reactivity to be used as a soil acidity corrector.

The efficiency of soil remediation depends on the value of the relative total neutralizing power (PRNT). The closer the PRNT is to

100, the quicker and more effective the corrections are in correcting soil acidity. Agrochemical analyses are necessary to see how much correction the soil needs (MARQUES *et al.,* 2006).

Brazilian legislation on fertilizers and correctives establishes the minimum standards for marketing these inputs. According to Benedini (1988), the relative total neutralizing power (PRNT) can be expressed by:

$$PRNT (\%) = PN \times RE / 100$$

Where:

PN - Total potential capacity of neutralizing bases contained in acidity correctives, expressed in pure calcium carbonate equivalent (% $CaCO_3$). The minimum PRNT value for commercialization is 45% and the PN is 67%.

RE - Particle reactivity. This value expresses the percentage (%) of the corrective agent that reacts in the soil within 12 to 36 months.

As for particle size, the minimum physical characteristics required are:

95% of particles mustpass throughsievesieve 2 mm sieve (ABNT - 10);

70% of particles mustpass throughsievesieve 0.84 mm sieve (ABNT - 20);

50% of theparticles mustpass throughsievesieve 0.30 mm sieve (ABNT - 50).

The reactivity of the corrective agent particles is calculated using the following formula: RE = 0*(ABNT10)+20*(ABNT10-20)+60*(ABNT20-50)+100*(ABNT50) 100

Where:

A. Zero reactivity (0%) for the fraction retained on the ABNT 10 sieve. These particles have no corrective effect over a period of 12 to 36 months;
B. Reactivity 20% for the fraction that passes the ABNT 10 sieve and is retained on the ABNT 20 sieve;
C. Reactivity 60% for the fraction that passes the ABNT 20 sieve and is retained on the ABNT 50 sieve;
D. 100% reactivity for the fraction that passes the ABNT 50 sieve.

According to Benedini (1988), limestones can be classified as follows:

Group A - PRNT of 45 - 60%
Group B - PRNT of 60.1 - 75% Group C - PRNT of 75.1 - 90% Group D - PRNT above 90.1%

The lower the soil pH (higher acidity), the less nutrients are used by plants; in the soil pH range of 6.0-6.5 (lower acidity), the availability of most nutrients is maximum. Some important nutrients for plant development, such as phosphorus, become limiting due to the acidity of the soil. It is estimated that the plant only takes advantage of 20 to 30% of the phosphorus applied in the form of chemical fertilizer (BRAGA, 2010).

Limestone can be classified according to its magnesium oxide (MgO) content as follows:

A) calcitics - less than 5%;

B) dolomitic - greater than or equal to 5%.

According to the pedological map of the State of São Paulo published by the Integrated Water Resources Management System of São Paulo (SIRGH-SP, 2010), in Aragatuba there is a predominance of Dark Red Latosol, which is generally highly weathered and has little mineral reserve for plants (OLIVEIRA, 1999).

On the other hand, they reach depths of more than 2 meters. Other characteristics are the high concentration of the clay fraction in the soil (15 to 80%) with low activity, providing low cation exchange capacity (CEC).

When the clay content is low, it has little capacity to store and retain water. The majority (95%) are dystrophic and acidic with a pH between 4 and 6. However, they have little slope, no more than 7%, making them suitable for mechanized farming. It should be noted that the soil's low fertility can be easily corrected with the application of fertilizers and correctives (KER, 1997).

The cation retention and adsorption capacity of the soil is measured using the cation exchange capacity (CEC) which is expressed in $cmol_c.dm^{(-3)}$, $mmol(_c).dm^{(-3)}$ or $mg.kg^{-1}$ depending on the region. The effective CEC is determined as a function of the soil pH and the potential CEC is determined using a buffer solution at pH 7.0. In acid soils, the effective CEC is lower than the potential CEC and is estimated by the sum of exchangeable bases, i.e. adsorbed ions such as Ca^{2+}, Mg^{2+}, K^+ , Na^+ , $H + Al^+$ 3 (BRAGA, 2010).

In the study by Cordeiro et al. (1988), it was observed that liming caused a 19% reduction in the availability of K^+ for sugarcane in purple latosol and 10% in red-yellow latosol.

Experiments carried out by Silveira et al. (1980) used four doses of K+ and two types of limestone in a dark red latosol and found that K+ responded at lower doses in the presence of calcitic limestone and at higher doses in dolomitic limestone, attributing this to the need to maintain an adequate K/(Ca + Mg) ratio. As for potassium (K^+), in sugarcane crops it responds intensely to its application (GAVA, 2008).

Table 1 shows the amount of limestone recommended for correcting the acidity of different types of soil, according to the pH level found after analysis:

Table 1 - Tons of limestone per hectare to be used to correct acidity

pH	Sandy land	Mixed land	Clay soils	Peaty lowlands
4,0	2,0	6,0	9,0	21,0
4,2	1,9	5,6	8,5	19,5
4,4	1,8	5,2	8,0	18,0
4,6	1,7	4,8	7,5	16,5
4,8	1,6	4,4	7,0	15,0
5,0	1,5	4,0	6,5	13,5
5,2	1,4	3,6	6,0	12,0
5,4	1,3	3,2	5,5	10,5
5,6	1,2	2,8	5,0	9,0
5,8	1,1	2,4	4,5	7,5
6,0	1,0	2,0	4,0	6,0

Source: MALAVOLTA, 2007

2.5 Ornamental plants

Ornamental plants are those that are distinguished by their flowering, the shape or color of the leaves and the overall shape or appearance of the plant. Ornamental plants form several groups in terms of the effect they can provide; their showy flowers and foliage. These groups are used together in beds, shade or full sun. They are used on their own or in individual pots, or are used to produce cut flowers (RIBEIRO, 2008).

Trees are larger, woody plants with a trunk, a defined canopy and an adult size of more than six meters. Their function is to provide ornamentation, shade, reduce the temperature range, guide and control winds, reduce noise and air pollution, attract and shelter birds and other small animals, helping to maintain the balance of nature.

The shrubs have woody characteristics, forking at a low height or close to the ground, with an adult size of less than six meters. As well as being ornamental, they delimit the view and guide the movement of people, providing privacy, complementing architectural lines, highlighting or hiding unsightly views, forming a vegetated curtain to protect against the sun, dust and noise.

Climbers can be classified as: voluptuous climbers when the stem has the habit of curling around a support in a spiral shape; sessile climbers (when the stems emit fixing organs, attaching the plants to the support with fixing roots, tendrils and hooks and lianas); climbers that don't have fixing organs, their branches initially grow upwards, then with the weight they bend downwards, forming an arch. From this arch comes a new sprout that repeats the cycle; and scandents (plants that take on a shrubby appearance when planted on their own, but when planted next to a support, they spread their branches and lengthen their stems in order to support themselves, as they have no fixing organs and need to be tied down in order to stay in the desired place).

Palms and cycads are plants of varying sizes with a characteristic appearance of both the trunk and the crown. Their trunks are called stipes and their typical leaves are usually pinnate and flabellate. They are types that impress mainly with their slender silhouette, characterizing a region's identity with their beautiful architectural lines.

Herbaceous plants are plants with non-woody or semi-woody stems (perennials or annuals) of varying size, which can acquire the height and effects of a shrub and can be planted in shady environments or entirely in the open. These ornamental plants replace shrubs in shady places, depending on the color and texture of their foliage or flowering, and serve as a contrast or eye-catcher.

Fodder plants are plants that grow horizontally significantly more than vertically, cover the surface of the ground and are distinctly different from lawns in that they are generally intolerant of direct sunlight and trampling. As well as ornamenting an environment, they protect the soil from erosion caused by wind and rain, and break up the monotony of extensive lawns, forming designs or emblems that increase the options for landscaping choices due to the varied textures and colors of their leaves.

Lawns are soil surfaces protected from weathering, formed exclusively by the grass family (Poaceae). These species are essential for the formation of courts for some sports.

Succulent plants generally live in arid regions or areas and have fleshy tissues that are very rich in water, providing a water reserve for periods of drought.

Aquatic plants differ from other plants in that they are: floating, when they have no fixation and are always on the surface of the water.

The production methods of ornamental plants require knowledge of the propagation methods of each species, the temperature characteristics in the seedling propagation phase are 20°C, in addition to high relative humidity, low insolation levels and phytosanitary control are very similar for different species (KAMPF, 2000).

3. METHODOLOGICAL PROCEDURES

The methodology used was a practical class held at Fundapao Casa (CASA Arapa) located in the city of Arapatuba in the state of Sao Paulo with the support of the Jose Augusto Lopes Borges State School.

Using the learning situation in the chemistry student handbook (vol. 1) in relation to the production and use of lime, the students were proposed to use the solid residue of school chalk to be used as an acidity corrector, according to the physical-chemical characteristics of the residue, in soil for ornamental plants, demonstrating a sustainable use of this material to the students.

Two types of school chalk used in the classroom were tested, plasticized (anti-allergenic chalk) and non-plasticized, to see which was the most effective material to use on the ground.

The student J. F. S. S. was assigned to collect the soil and prepare the school chalk waste to start the practical and demonstrate the procedures and results of the practical lesson to the other students.

3.1. Collecting soil for the practical class

The soil was collected in October 2015 at the Fundapao Casa (CASA Arapa) in the municipality of Arapatuba, in the northwestern region of the state of São Paulo, for acidity analysis. The methodology was based on Embrapa Solos (EMBRAPA, 1997), for taking acidity (pH) readings.

Before the practical class, the procedure for collecting and preparing the materials to be used in the classroom was carried out.

Figure 1 shows the student J. F. S. S. collecting soil to carry out the acidity analysis for the practical lesson to be held in the Chemistry class of the 1st Year of High School.

Figure 1 - Soil collection for acidity reading (pH)

Source: Author's archive

3.2. Preparation and weighing of soil for acidity analysis

After collection, the soil was weighed on an analytical balance and sieved in order to obtain a smaller particle size for acidity (pH) analysis. Figure 2 shows the student weighing and sieving the soil.

Figure 2 - Weighing and sieving the soil.

Source: Author's archive

3.3. Weighing the sieved soil

After the sieving process, the samples were weighed again to obtain the right soil for the acidity analysis (pH).

In order to carry out the analysis, one pot was separated from the control soil and two pots were treated with ground school chalk.

The student carried out the process shown in figure 3, weighing the sieved soil and separating the pots for analysis and demonstration.

Figure 3 - Weighing the sieved soil and separating the pots for analysis.

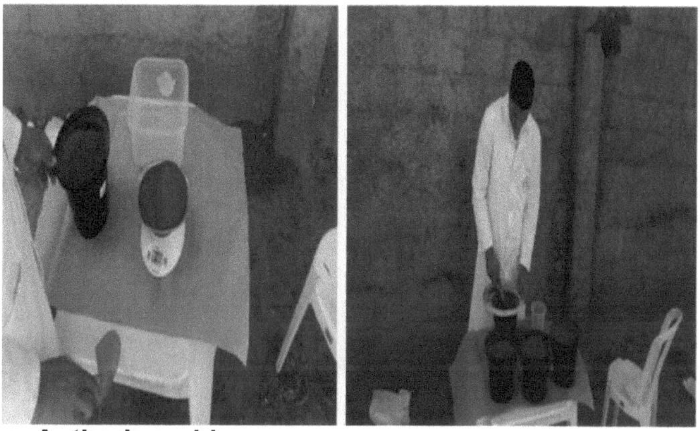

Source: Author's archive

3.4. Reactivity of plasticized and unplasticized chalk

The student demonstrated the reactivity of plasticized school chalk and non-plasticized chalk in acetic acid (4%) to the rest of his classmates, in order to find out which chalk is suitable for neutralizing acidity.

Approximate pH readings were taken on universal acidity (pH) indicator paper to find out which chalk (plasticized or non-plasticized) was more reactive.

In the classroom with the other students, the reactivity test was added to the teaching of the learning situation so that the students could see the importance of using the solid residue in such a way as to ensure that it is used more efficiently in the soil. Figure 4 shows the reactivity tests of the chalk in acetic acid and the reading of the pH results.

Figure 4 - Reactivity analysis of plasticized and non-plasticized school chalk

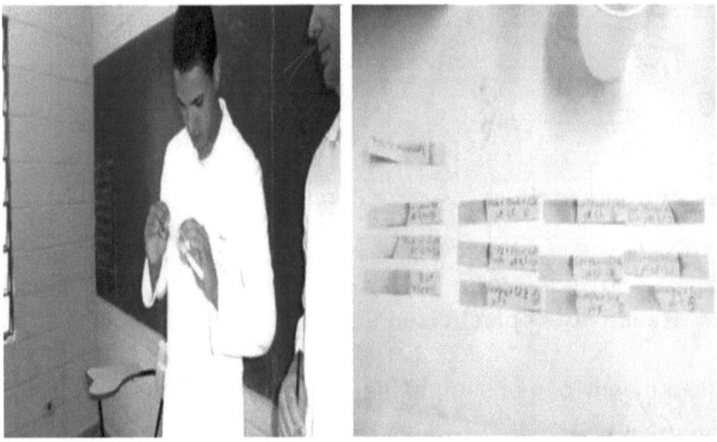

Source: Author's archive

3.5. Macerapao school chalk

With the results observed in the classroom, the choice was made to macerate unplasticized school chalk. The student macerated the chalk, then sieved it to obtain the smallest possible particle size so that the chemical reaction could take place properly to neutralize the acidity of the soil. Figure 5 shows the maceration of the school chalk to be used on the collected soil.

Figure 5 - Macerating school chalk for use as a soil improver.

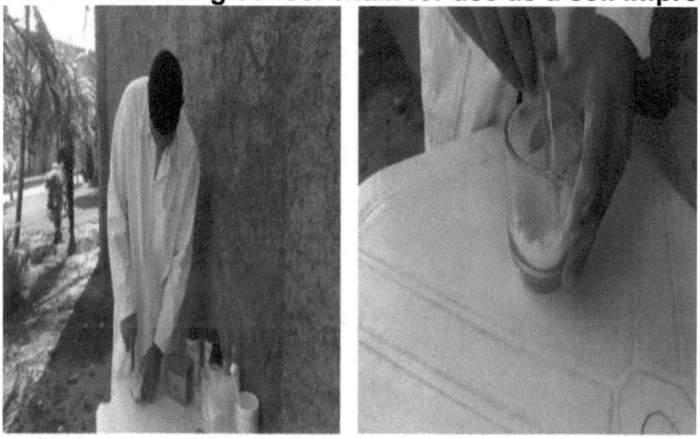

Source: Author's archive

3.6. Weighing the sample of ground school chalk residue.

The macerated school chalk was weighed on a digital scale and the amount of macerated material was sieved to reduce the particle size for the soil neutralization reaction. Figure 6 shows the weighing of the macerated school chalk.

Figure 6 - Weighing the macerated school chalk.

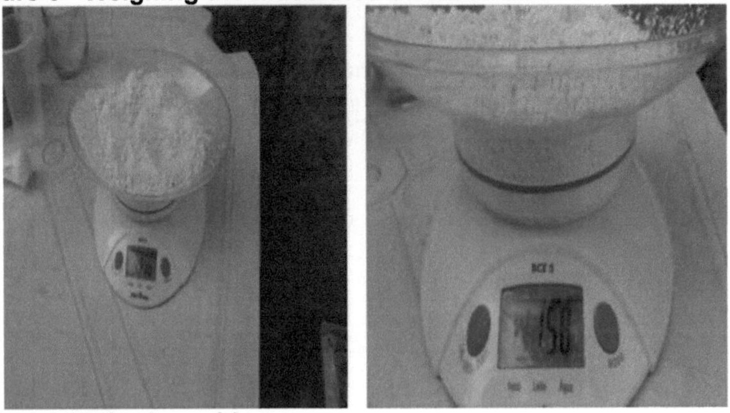

Source: Author's archive

3.7. Application of ground school chalk waste in ornamental planting pots

Ground school chalk waste was applied to the pots with sieved soil to be analyzed for soil acidity neutralization. Figure 7 shows the application of ground school chalk waste in pots for planting ornamental plants.

Figure 7 - Application of ground school chalk residue to pots for ornamental plants.

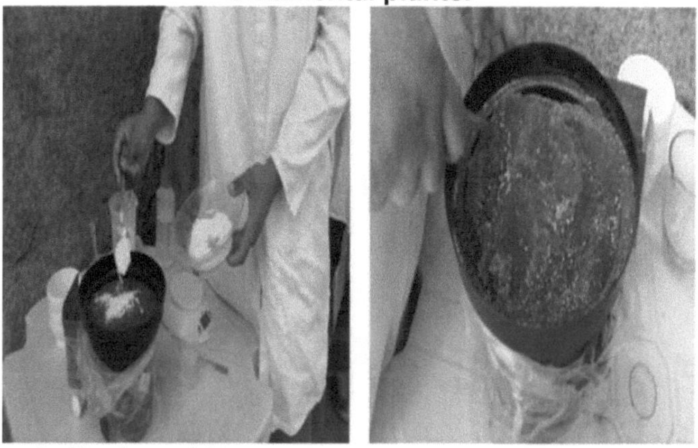

Source: Author's archive

3.8. Reaction time and filtering of the control soil and soil treated with ground school chalk waste for pH analysis.

A sample was taken from the control pot and from the pots treated with ground school chalk waste to measure the pH in calcium chloride ($CaCl_2$) according to the recommended methodology (EMBRAPA, 1997).

Soil acidity (pH) was analyzed after filtering the samples with calcium chloride. The pH was read using a Kasvi® portable pH meter (K39-0014P) for the comparative study. Figure 8 shows the filtering of the soil in a solution of calcium chloride ($CaCl_2$) for the pH reading of the samples.

Figure 8 - Samples of control soil and soil treated with ground School chalk for pH reading

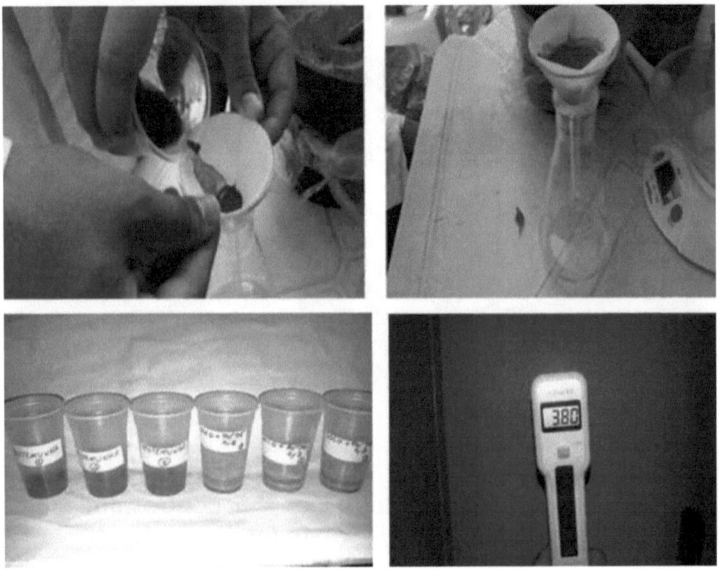

Source: Author's archive

4. RESULTS AND DISCUSSION

The tests indicated that the unplasticized school chalk acted with greater efficiency in raising the pH of the acetic acid solution (4%), *in an in vitro* test, from 4.0 to pH 5.5 in an interval of 15 (fifteen) minutes.

The unplasticized chalk was manually macerated and applied to three pots in a ratio of 1:100 (g) for analysis in calcium chloride solution ($CaCl_2$) 0.01(Mol-$L^{-1)}$. A Kasvi® portable pH meter (K39-0014P) was used to check the results. The pH of the soil went from 5.56 to 6.75, as shown in Table 1.

Table 1. Methodology for reading pH in Calcium Chloride

Soil pH analysis (untreated) in Calcium Chloride ($CaCl_2$)	**pH analysis of soil with ground school chalk in calcium chloride** (CaCl2)
Witness 1: pH: 5.5	1. Soil and school chalk residue: pH: 6.7
Witness 2: pH: 5.6	2. Soil and school chalk residue: pH: 6.8
Witness 3: pH: 5.5	3. Soil and school chalk residue: pH: 6.7
Arithmetic mean: pH: 5.56	**Arithmetic mean: pH: 6.75**

Source: Structured by the author

Through the results shown on the portable pH meter, the students were able to see the change in the acidity level compared to the control soil (untreated) and understand the importance of using lime in agricultural processes in the teaching/learning process, using the solid residue of school chalk.

With these results in hand, the learning situation on the use and production of lime for the students and the use of solid waste, demonstrated a greater interest in conservation and rethinking the importance of using waste in a conscious and planned way.

5. FINAL CONSIDERATIONS

Environmental education brings perspectives so that the natural and social environments interact in harmony. As agents of social and environmental change, students must associate the learning situation with the new environmental paradigms the world is facing. According to the National Curriculum Parameters (PCN), student development must also be associated with environmental awareness.

Holding practical classes for environmental education at Fundapao Casa (CASA Arapa) establishes new perspectives in the teaching/learning process. In addition to the involvement of the entire management team, teachers and students in the search for awareness of the reuse of waste, valuing learning.

Ground school chalk waste proved to be efficient in contextualizing the situation of learning how to produce and use lime to correct soil acidity, demonstrating that a widely discarded material can be used in pots with ornamental plants, providing a sustainable alternative to disposing of this waste.

6. BIBLIOGRAPHICAL REFERENCES

ABRANTES, V. A. M. S.; MIRANDA, M. G.; VASCONCELLOS, C. A. B. Educapao ambiental na escola: do papel a realidade - do discurso ao comprometimento. **Revista Brasileira de Educapao Ambiental.** v.9, n. 2 p. 424, 435, 2014.

ABRELPE. **Panorama of Solid Waste in Brazil**. 2014. Available at: <http://www.abrelpe.org.br/Panorama/panorama2014.pdf>. Accessed on: July 6, 2016.

ALCARDE, J. C.; RODELLA, A. A. **Topicos em ciencia do solo In: Qualidade e Legislapao de fertilizante e corretivos**. Vol. 3, ed 3. SBCS p. 291-334, 2003.

AMBIENTE BRASIL. **Educapao Ambiental.** 2008. Available at: <www.ambientebrasil.com>. Accessed on: July 4, 2016.

ARAUJO, T. C. A. **Main world historical milestones in** environmental **education**. 2007. Available at: <http://noticias.ambientebrasil.com.br/artigos/2007/09/11/33350-principais- marcos-historicos-mundiais-da-educacao-ambiental.html>. Accessed on: July 6, 2016.

BENEDINI, M. S. **New concept in the use of limestone in sugarcane**. COPERSUCAR, Serie Agronomica, n. 16, ed. 1, 1988.

BITTENCOURT, W.C.; SAKAI, M. Leaching of native K^{+1} from tropical soils. **Piracicaba: Cena**, 1975.

BRAGA, G. N. M. **Calagem - método para determinar a necessidade de calcario**. Available at: <http://agronomiacomgismonti.blogspot.com/2010/01/calagem-metodos- para-determinar.html>. Accessed on: July 20, 2016.

BRAZIL. **Environmental Education: sustainability apprentices**. 2007. Available at: <http://portal.mec.gov.br/dmdocuments/publicacao2.pdf>. Accessed on July 6, 2016.

CORDEIRO, D. A.; BATISTA, L. F.; GURGEL, M. N.; BITTENCOURT, V. C. **Study by means of labeling techniques on the K-liming relation in soils cultivated with sugarcane**. In: Congress of The International Society of Sugar Cane Technologists. Sao Paulo: 1988.

EL-DEIR, S. G. **Solid waste: perspectives and challenges for integrated management.** 2014. Available at: <file:///C:/Users/Andrea%20Chiarioni/Downloads/veja_aqui-_ebook_residuos_solidos.pdf>. Accessed on: July 6, 2016.

EMBRAPA. **Manual of Soil Analysis Methods.** 2.ed. Rio de Janeiro, 1997.

GALLI, A. **Environmental education as a tool for sustainable development**. 2007. Master's dissertation in Economic and Social Law. Pontifical Catholic University of Paraná, 2007.

GAVA, R. **Effect of vinasse application with nitrogen addition on the growth of sugarcane *(Saccharum officinarum* L.)**. 53f. Monograph (Graduation in Agricultural Engineering), State University of Maringa (UEM). Cidade Gaucha, Parana, 2008.

GIZMODO BRASIL. **Actually, the chalk used on blackboards isn't exactly chalk.** 2015. Available at: <http://gizmodo.uol.com.br/na-verdade-o-giz-usado-nas-lousas-nao-e-exatamente-giz/>. Accessed on: 06 Jul. 2016.

GODECKE, M. V.; NAIME, R. H.; FIGUEIREDO, J. A. S. Consumerism and the generation of urban solid waste in Brazil. **Revista Eletronica: Gestao, Educagao e Tecnologia Ambiental**. v. 8 n. 8 p. 1700-1712, Sep./Dec. 2012.

GOMES, J. I. **Utilization of Wood Waste in Three Timber Companies in the State of Para**. Embrapa Technical Communication, 2004.

IMPLASTEC. **Graphite powder: basic characteristics**. 2011. Available at: <http://implastec.com.br/revenda-eletronica/grafitec-em-po/>. Accessed on: September 2, 2017.

INEP - NATIONAL INSTITUTE FOR EDUCATIONAL STUDIES AND

RESEARCH ANISIO TEIXEIRA. **Inclusion of Science in Saeb: basic document.** Brasilia: Instituto Nacional de Estudos e Pesquisas Educacionais Anisio Teixeira, 2013.

JANKE, N. **National environmental education policy: contradictions and disputes.** Bauru, 2012. Thesis (Doctorate in Science Education) - Faculty of Sciences, São Paulo State University - UNESP.

JACOBI, P. **Educapao ambiental, cidadania e sustentabilidade.** 2003. Available at: <http://www.scielo.br/pdf/cp/n118/16834.pdf>. Accessed on: July 4, 2016.

KAMPF, A. N.; FERMINO, M. H. **Encontro nacional sobre substrato para plantas: substrato para plantas: a base da produpao vegetal em recipientes**. Porto Alegre: Genesis, 2000.

KER, J. C. Latosols of Brazil: a review. **Revista Geonomos**. v. 5, n. 1, p. 17-40, 1997.

MALAVOLTA, E. **Sugarcane nutrition and fertilization**. 2007. Available at: <<http://www.malavolta.com.br/pdf/730.pdf>. Accessed on: July 17, 2016.

MARQUES, M. O.; MUTTON, M. A.; AZANIA, A. A. P. M.; TASSO JUNIOR, L. C.; NOGUEIRA, G. A.; VALE, D. W. **Topicos em tecnologia sucroalcooleira**. Ed. Marcos Omir, 2006.

MAZZER, C.; CAVALCANTI, O. A. Introduction to environmental waste management. **Revista: Infarma**, v. 16 p. 67, 77, 2004.

CREATIVE MIND. **Graphite powder - what's it for?** 2010. Available at: <https://amentecriativa.wordpress.com/2014/11/25/grafite-em-po-para-que- serve/>. Accessed on: September 2, 2017.

MMA - MINISTRY OF THE ENVIRONMENT. **Organic Waste Management**.
2016. Available at: <http://www.mma.gov.br/cidades-sustentaveis/residuos-solidos/gest%C3%A3o-de-res%C3%ADduos-org%C3%A2nicos>. Accessed on: July 20, 2016.

MMA - MINISTRY OF THE ENVIRONMENT. **The different shades of environmental education in Brazil.** 2009. Available at: <http://www.mma.gov.br/estruturas/educamb/_arquivos/dif_matizes.pdf> Accessed on: 06 Jul. 2016.

MMA - MINISTRY OF THE ENVIRONMENT. **National solid waste plan.** 2012. Disponivel em: <http://www.sinir.gov.br/documents/10180/12308/PNRS_Revisao_Decreto_2 80812.pdf/e183f0e7-5255-4544-b9fd-15fc779a3657>. Accessed on: July 6, 2016.

OLIVEIRA, J.B. **Soils of the State of São Paulo: description of the classes recorded on the pedological map**. Boletim Cientifico, n. 45. Campinas, Sao Paulo, 1999.

OLIVEIRA, I. P.; COSTA, K. A. P.; RODRIGUES, C. MACEDO, F. R.; MOREIRA, F. P.; SANTOS, K. J. G. Maintenance and correction of soil fertility for the inclusion of the cerrado in the production process. **Revista Eletronica Faculdade Montes Belos - Goias**, v.1, n.1, p. 50-64, Montes Belos, Goias 2005.

NATIONAL CURRICULUM PARAMETERS. **Introduction to the national curriculum parameters.** 1997. Available at: <http://portal.mec.gov.br/seb/arquivos/pdf/livro01.pdf>. Accessed on: July 6, 2016.

PEREZ, D. V. **Química na agricultura**. 2007. Available at: < http://web.ccead.puc-rio.br/condigital/mvsl/Sala%20de%20Leitura/conteudos/SL_quimica_na_agric ultura.pdf>. Accessed on: July 14, 2016.

PRS - SOLID WASTE PORTAL. **A brief history of global environmental education**. 2013. Available at: <http://www.portalresiduossolidos.com/historia-da-educacao-ambiental- brasil-e-mundo/>. Accessed on: July 6, 2016.

RAYMUNDO, V.; NEVES, M. A.; CARDSO, M. S. N.; BREGONZI, I. S.;
LIMA, J. S. S.; FONSECA, A. B. Residues of marble sawdust as a soil acidity corrector. **Revista Brasileira de Engenharia Agricola e Ambiental**, v.17, n.1, p.47-53, 2013.

CONAMA RESOLUTION. **Conama Resolution No. 5.** 1993. Available at: <http://www.mma.gov.br/port/conama/legiabre.cfm?codlegi=130>. Accessed on: July 4, 2016.

RIBEIRO, J. C. D. **Ornamental plants: an alternative for the development of family farming in the canoas settlement project, in the municipality of Presidente Figueiredo**. Master's dissertation in Agricultural Sciences, National Institute for Amazonian Research - INPA, Federal University of Amazonas - UFAM, 2008.

ROSA, A. B. M.; MATOS, G. F.; BATISTA, V. M. S. **Waste management: reuse of graphite from school pencils as a lubricant for hinges.** In: X Congresso de Iniciapao Cientifica Junior do Centro Universitario de Adamantina, Sao Paulo, 2017.

ROSSETTO, R.; SPIRONELLO, A.; CANTARELLA, H.; QUAGGIO, J.A. Liming for sugarcane and its interaction with doses of Potassium. **Bragantia**, v. 63, n. 1, p. 105-119, Campinas, SP, 2004.
ROSSETO, R.; FERREIRA. D. F. L.; VITTI. A. C. **Fertilidade do solo, nutripao e adubapao**. v.1 p. 221-229, 2008.

SAO PAULO. **Lime production and use**. Available at: <http://pt.slideshare.net/diogo087/caderno-do-aluno-qumica-1-ano-vol-1- 2014-2017>. Accessed on: 06 jul. 2016.

SCHALCH, V. LEITE, W. C. A.; FERNANDES JUNIOR, J. L.; CASTRO, M.
C. A. A. **Solid waste management**. 2002. Available at: <http://www.deecc.ufc.br/Download/Gestao_de_Residuos_Solidos_PGTGA/
Apostila_Gestao_e_Gerenciamento_de_RS_Schalch_et_al.pdf>. Accessed on: July 21, 2017.

SILVEIRA, J. F.; SIQUEIRA, J. O.; GUEDES, G.A. A. **Interapao fosforo x potassio x calcario em cana de-apucar (cana planta)**. Brasil Apucareiro, Rio de Janeiro, v.95, n.1, p.18-21, 1980.

SILVA, A. P. M.; VIANA, J. P.; CAVALCANTI. A. L. B. **Solid waste from mining activities.** 2011. Available at: <http://www.cnrh.gov.br/projetos/pnrs/documentos/cadernos/11_CADDIAG_ Res_Sol_Mineracao.pdf>. Accessed on: July 6, 2016.

SILVA, J.O. **Perfil da Minerapao do Calcario (Technical Report No. 38 of the Energy Sector Technical Assistance Project - ESTAL Project)**.
2009. Disponivel em:
<http://www.mme.gov.br/;jsessionid=5DB4322472C47AA67B0A704B 635A61 3C.srv155>. Accessed on: July 6, 2016.

SIRGH - SP **Integrated Water Resources Management System of Sao Paulo. Pedological map of Arapatuba - SP**. Available at: <http://www.sigrh.sp.gov.br/sigrh/ARQS/RELATORIO/CBH-BT/543/rel_zero/pedologico.pdf>. Accessed on: July 10, 2016.

QUITINO, C. A. A. A **history of environmental education in Brazil and worldwide.** 2013. Available at: <http://www3.unifai.edu.br/pesquisa/publica%C3%A7%C3%B5es/artigos- cient%C3%ADficos/professores/bacharelados/um-histist%C3%B3rico-sobre- educa%C3%A7%C3%A3o>. Accessed on: July 4, 2016.

WANGEN, D. R. B.; FREITAS, I. C. V. Domestic composting: an alternative way of using solid organic waste. **Revista Brasileira de Agroecologia.** 5(2): 81-88, 2010.

MIX
Papier aus verantwortungsvollen Quellen
Paper from responsible sources
FSC® C105338

Printed by Books on Demand GmbH, Norderstedt / Germany